Contents

Some words are shown in bold, **like this**. You can find out what they mean by looking in the Glossary.

Numerals

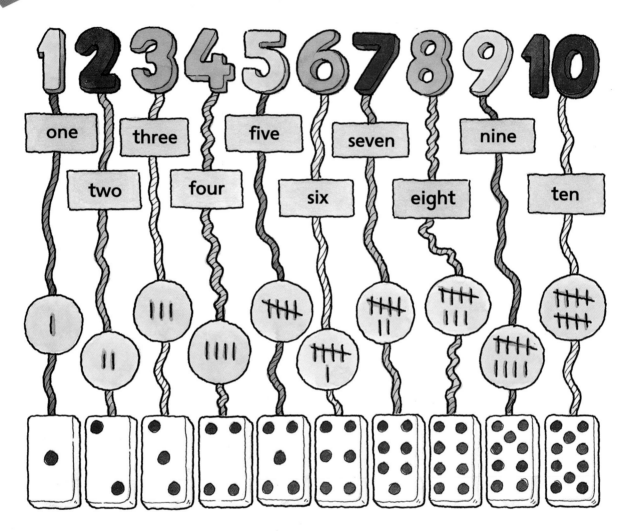

Numbers can be shown in lots of different ways.

We can use words, pictures, signs or **symbols**.

They are all called **numerals**.

Numbers

Peter Patilla

First published in Great Britain by Heinemann Library,
Halley Court, Jordan Hill, Oxford OX2 8EJ,
a division of Reed Educational and Professional Publishing Ltd.
Heinemann is a registered trademark of Reed Educational & Professional Publishing Limited.

OXFORD MELBOURNE AUCKLAND
JOHANNESBURG BLANTYRE GABORONE
IBADAN PORTSMOUTH NH (USA) CHICAGO

Designed by AMR
Illustrations by Art Construction and Jessica Stockham (Beehive Illustration)
Originated by HBM Print Ltd, Singapore
Printed and bound by South China Printing Co., Hong Kong/China

04 03 02 01 00
10 9 8 7 6 5 4 3 2 1

ISBN 0 431 09359 8
This title is also available in a hardback library edition (ISBN 0 431 09352 0)

British Library Cataloguing in Publication Data
Patilla, Peter
  Numbers. – (Maths links)
  1.Numeration – Juvenile literature
  I.Title.
  513.2

Acknowledgements
The Publishers would like to thank the following for permission to reproduce photographs:
Allsport, pgs 11 /Craig Prentis; Trevor Clifford, pgs 5, 7, 9, 10, 12, 13, 14, 15, 17, 18, 19, 20, 23, 24, 27, 28; Tony Stone Images, pgs 29 (t) /Paul Chesley, (b) /Jon Riley.

Cover photograph reproduced with permission of Trevor Clifford.

Our thanks to David Kirkby for his comments in the preparation of this book.

Every effort has been made to contact copyright holders of any material reproduced in this book. Any omissions will be rectified in subsequent printings if notice is given to the Publisher.

For more information about Heinemann Library books, or to order, please phone +44 (0)1865 888066, or send a fax to +44 (0)1865 314091. You can visit our website at www.heinemann.co.uk

All sorts of numerals can be used to show numbers. Here are some of those you might see.

Look around you and find some numerals. Where did you find them?

Digits

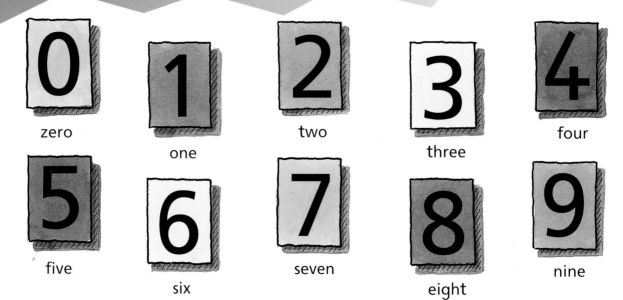

zero · one · two · three · four

five · six · seven · eight · nine

These are the digits and their names.

These numbers use 2 digits.   These numbers use 3 digits.

There are 10 **digits**. They are the numbers 0, 1, 2, 3, 4, 5, 6, 7, 8 and 9. Digits are very important numbers because they are used to build up other numbers.

Examples of digits.

Digits are used to write or print all sorts of numbers. Big numbers, little numbers, money and measurements all use digits.

Can you find an example of each digit in the numbers you see around you?

Ordering numbers

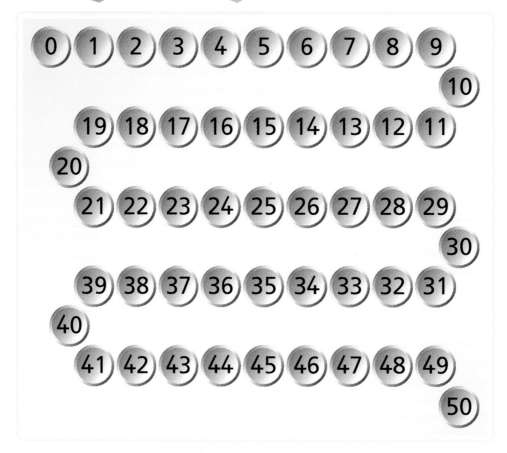

The numbers 0 to 50 in order.

Numbers have an order. The order can be forwards or backwards. Numbers can go on and on and never stop. It is important to know the order of numbers.

8

We see numbers arranged in order in all sorts of places. You can see the numbers 1 to 12 in order on some clock faces.

Where can you see numbers written in order in the picture?

First to last

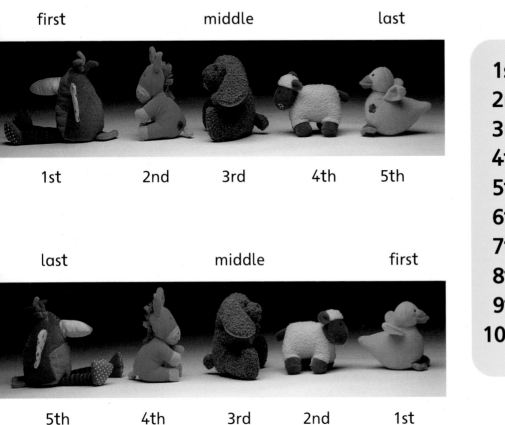

first   middle   last

1st   2nd   3rd   4th   5th

last   middle   first

5th   4th   3rd   2nd   1st

1st	first
2nd	second
3rd	third
4th	fourth
5th	fifth
6th	sixth
7th	seventh
8th	eighth
9th	ninth
10th	tenth

When a line turns round, first becomes last.

When things are in a line, we use words like first, second and third. Five things go from first to fifth. Ten things go from first to tenth.

Order is important. It is often nice to be at the front of a queue, rather than the back.

Look at the picture. Who is first? Who is second? Who is third? Who is fourth, and last?

Sets

Things are often packed in twos, threes, sixes and twelves. Sometimes words such as **pair**, **dozen** and **half dozen** are used. These words also tell us how many are in the set.

Quantities of things are sometimes put into patterns. These patterns help us to count. They also help to show when some are missing.

Look at the picture. How many are missing?

Comparing sets

more   fewer   equal

more   less   equal

8 > 6 means 8 is more than 6
6 < 8 means 6 is less than 8
1 and 1 = 2 means 1 and 1 equals 2

When we compare two amounts or numbers, we can use words such as **more**, **fewer**, **less**, **equal** and same. We can also use the signs **>** **<** and **=**.

14

Sometimes we like to compare who has the **most** and who has the **least**.

Which plate has the most sandwiches? Which plate has fewest crisps? Which glass has the least juice?

Addition

+

= 9

adding two sets

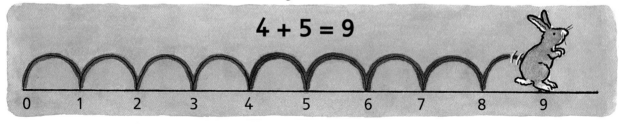

adding three sets

4 + 5 = 9

0 1 2 3 4 5 6 7 8 9

jumping along a number line

Adding is putting two or more sets together.
It is also jumping forward along a number line.
The answer to an **addition** is called the **sum**
or **total**. So the sum of 4 + 5 is 9.

In this game, the corner numbers add up to the number in the middle of the triangle.

Which numbers are hidden by the children's fingers? Make your own triangle totals cards and play the game.

Subtraction

taking away

5 take away 2 leaves 3
5 − 2 = 3

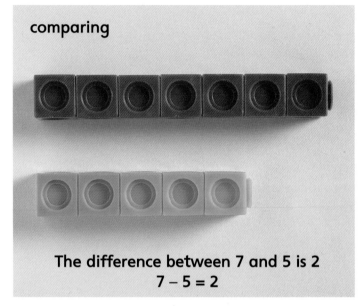

comparing

The difference between 7 and 5 is 2
7 − 5 = 2

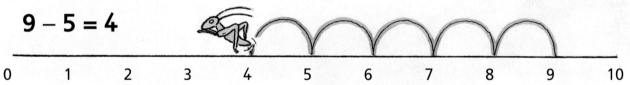

9 − 5 = 4

0 1 2 3 4 5 6 7 8 9 10

jumping along a number line

Subtraction (−) is taking away one number from another. It is also comparing two numbers to find the **difference** between them. Subtraction is also jumping backwards along a number line.

18

Each child has 10 beads and is hiding some of them.

How many beads is each child showing? How many is each child hiding? Why not play this subtraction game with someone?

Multiplication

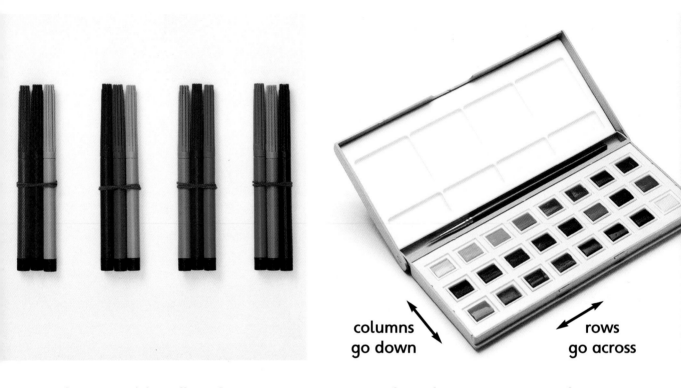

columns
go down

rows
go across

There are 4 bundles of pens.
Each bundle has 3 pens.
3+3+3+3=12

The colours are arranged in
rows and columns.

Multiplication (x) is adding the same number lots
of times. Sometimes the numbers to be added are
bundled together into sets. At other times the
numbers are put into rows and columns.

When things are put into rows and columns, it makes counting easier. Counting in twos, or threes, or fours, or fives saves having to count in ones.

How many biscuits are on the tray?

Sharing

9 cherries

3 equal sets: $9 \div 3 = 3$

2 equal sets with a remainder:
$9 \div 4 = 2$, remainder 1

Division (÷) is sharing something into equal sets or equal amounts. When dividing fairly there is sometimes a **remainder**.

22

We sometimes arrange things into small groups, such as twos and threes.

Which of these sets can be put into:
  twos without leaving a remainder?
  threes without leaving a remainder?

23

Odd and even

Even numbers

6 socks make 3 pairs.
6 is an even number.

Odd numbers

9 socks make 4 pairs,
with 1 left over.
9 is an odd number.

even numbers: 2, 4, 6, 8, 10, 12, 14, 16
odd numbers   1, 3, 5, 7, 9, 11, 13, 15

Even numbers can be put into twos exactly.
Odd numbers always have one left over when put
into twos. **Zero** means no number is present, so it
cannot be odd or even.

Look at the last digit of large numbers.

If it is 0, 2, 4, 6, 8, then the number is even.

If it is 1, 3, 5, 7, 9, then the number is odd.

Look at the numbers in the picture. Which numbers are odd, and which are even?

Fractions

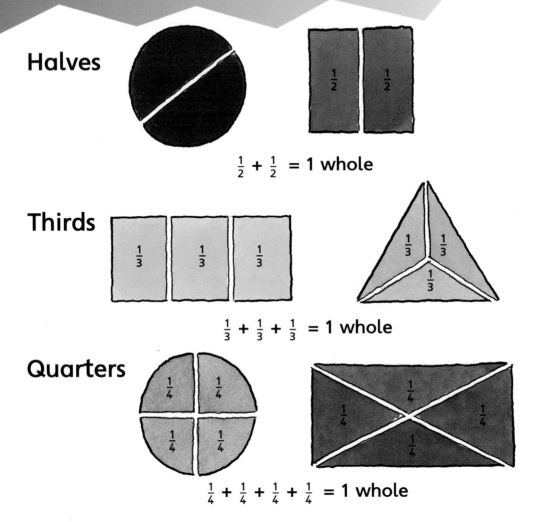

Halves

$\frac{1}{2} + \frac{1}{2}$ = 1 whole

Thirds

$\frac{1}{3} + \frac{1}{3} + \frac{1}{3}$ = 1 whole

Quarters

$\frac{1}{4} + \frac{1}{4} + \frac{1}{4} + \frac{1}{4}$ = 1 whole

We can divide shapes and quantities into **fractions**. Fractions are parts of something. When we halve something, each part must be the same size. This is the same for fractions such as thirds and quarters.

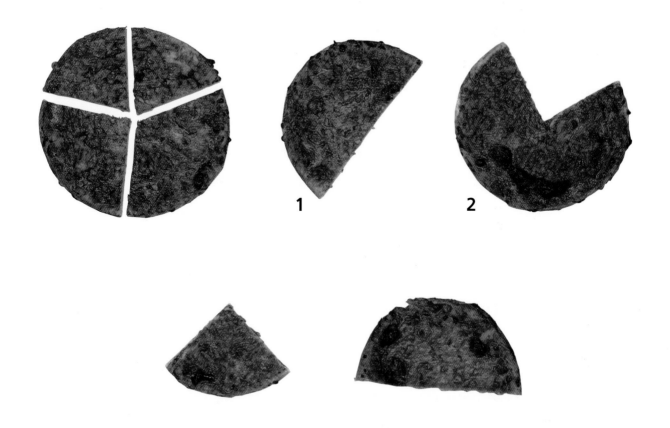

When you cut something up fairly, each fraction is the same size.

Is the whole pizza divided into equal quarters? Which fraction is missing from pizza 1? Which fraction is missing from pizza 2?

Large numbers

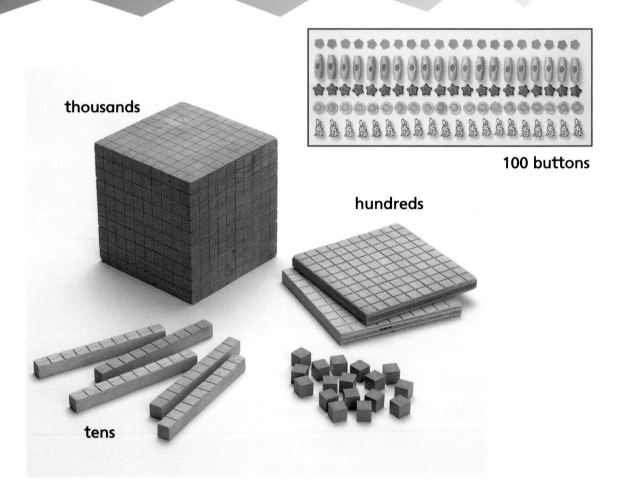

100 buttons

thousands

hundreds

tens

When we describe large numbers, we use words like **hundreds, thousands,** tens of thousands and **millions.** Trying to imagine large numbers is not very easy.

We find large numbers of people in schools,
at sports meetings and on busy streets.

Which picture shows thousands of people?

Glossary

addition (+) putting sets together and counting how many are in the combined set. See **plus**.

difference how many more one number is than another

digits all the numbers 0, 1, 2, 3, 4, 5, 6, 7, 8, and 9. They are used to write other numbers.

division (÷) sharing something into equal sets. It is also finding how many equal sets are in a number. Some divisions leave a remainder.

dozen 12 of anything

equal (=) used when things are worth the same

fewer when there are not as many

fraction part of a whole. Halves, thirds and quarters are examples of fractions.

half dozen 6 of anything

hundred (100) ten tens or a hundred ones. Hundreds numbers have 3 digits, for example 486.

least when something is fewer or has less than the rest

less (<) when there is not as much of something

million (1 000 000) a thousand thousands. Millions numbers have 7 digits, as in 2 550 000.

more (>) when there is a larger amount of something

most when something has more than the rest

minus (–) this means take away in subtraction

multiplication (×) a way of adding the same number lots of times, so 3×4 is the same as $4 + 4 + 4$.

numerals the ways in which we write and show numbers. They can be words, pictures, symbols or numbers.

pair when things are put into twos

plus (+) this means add together in addition

remainder what is left over after you have taken away, shared or divided something

subtraction (–) taking one set from another and finding what is left. See **minus**.

sum the answer we get when we add numbers together (addition). It is sometimes used to describe problems such as subtraction, multiplication and division.

symbols signs which stand for a word or something to do. The symbols for add, subtract, multiply and divide are + – × and ÷

thousand (1000) ten hundreds or a hundred tens. Thousands numbers have 4 digits, as in 3250

total the answer to an addition problem

zero (0) one of the digits. It stands for an empty set, or something with nothing in it.

Answers

page 9   telephone, clock, calculator, calendar

page 11   1st = 275   2nd = 679   3rd = 233   4th = 414

page 13   4 missing

page 15   plate on right; plate on right; glass in front

page 17   girl: 5, boy: 8

page 19   the boy is showing 4; he is hiding 6
                the girl is showing 6; she is hiding 4

page 21   12 (3 x 4)

page 23   twos: notebooks, pencils; threes: rubber bands,
                pencils. The pencils can be put into twos or threes.

page 25   odd: 49, 57, 31, 95, 15, 163
                even: 62, 42, 38, 40

page 27   no
                pizza 1: missing one half
                pizza 2: missing one quarter

page 29   top picture

Index